MATHWORKS!

Using Math to Build a SKYSCRAPER

by Hilary Koll, Steve Mills,
and William Baker

Math and Curriculum Consultant:
Debra Voege, Science and Math
Curriculum Resource Teacher

GARETH**STEVENS**
GS
PUBLISHING
A Member of the WRC Media Family of Companies

Please visit our web site at: **www.garethstevens.com**
For a free color catalog describing Gareth Stevens Publishing's
list of high-quality books and multimedia programs, call
1-800-542-2595 (USA) or 1-800-387-3178 (Canada).
Gareth Stevens Publishing's fax: (414) 332-3567.

Library of Congress Cataloging-in-Publication Data

Koll, Hilary.
    Using math to build a skyscraper / Hilary Koll, Steve Mills,
and William Baker. — North American ed.
        p. cm. — (Mathworks!)
    ISBN-10: 0-8368-6764-5 — ISBN-13: 978-0-8368-6764-0 (lib. bdg.)
    ISBN-10: 0-8368-6771-8 — ISBN-13: 978-0-8368-6771-8 (softcover)
    1. Mathematics—Problems, exercises, etc.—Juvenile literature.
2. Skyscrapers—Design and construction—Juvenile literature.
I. Mills, Steve, 1955- II. Baker, William, 1953- III. Title. IV. Series.
QA43.K623   2006
510.76—dc22   2006045010

This North American edition first published in 2007 by
Gareth Stevens Publishing
A Member of the WRC Media Family of Companies
330 West Olive Street, Suite 100
Milwaukee, Wisconsin 53212

This U.S. edition copyright © 2007 by Gareth Stevens, Inc.
Original edition copyright © 2006 by ticktock Entertainment
Ltd. First published in Great Britain in 2006 by ticktock Media
Ltd., Unit 2, Orchard Business Centre, North Farm Road,
Tunbridge Wells, Kent, TN2 3XF, United Kingdom.

Technical Consultant: William Baker
As an architect and structural engineer, Bill has developed
the structural systems of some of the world's tallest buildings,
including the 160-story Burj Dubai.

Gareth Stevens Editor: Dorothy L. Gibbs
Gareth Stevens Art Direction: Tammy West

Photo credits (t=top, b=bottom, c=center, l=left, r=right)
Shutterstock: 6tr (Norman Chan), 6br (Yan Vugenfirer), 8-9
(Jonathan Pais), 10-11 (Leah-Anne Thompson), 12-13 (luke
james ritchie), 14-15 (Jim Jurica), 16-17 (Robert Cumming),
18-19 (Wendy Kaveney Photography), 22-23 (Ismael Montero
Verdu), 24-25 (Albo), 26cr (Wen-ho Yang). Photodisc: 6-7,
26-27. Corbis: 20cr. FEMA: 20-21 (Robert A. Eplett).

Printed in the United States of America

1 2 3 4 5 6 7 8 9 10 09 08 07 06

CONTENTS

HAVE FUN WITH MATH

How to Use This Book

Math is important in the daily lives of people everywhere. We use math when we play games, ride bicycles, or go shopping, and everyone uses math at work. Imagine you have been hired to build a skyscraper! You may not realize it, but architects and structural engineers use math for everything from the first design to the final brick. In this book, you will be able to try lots of exciting math activities, using real-life data and facts about tall buildings. If you can work with numbers, measurements, shapes, charts, and diagrams, then you could BUILD A SKYSCRAPER.

How does it feel to build a tower?

Grab your blueprints and find out what it takes to design and construct one of the world's tallest buildings.

THE BUILDING SITE

To design a skyscraper, you must first know the size and shape of the building site, which is the plot of land you will build on. The dimensions of the plot will help you decide the size and shape of your building. Many cities have rules about how large a building can be constructed on a certain size building plot. In some cities, the size of a building has to be based on a formula known as Floor Area Ratio (FAR). In cities that do not use FAR, the size of a building usually must be approved by city planners before construction can begin. After choosing a suitable building site, you are ready to begin your design.

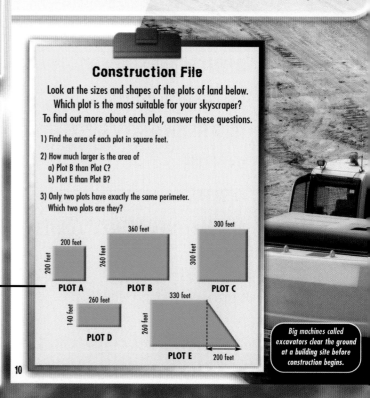

Construction File

Look at the sizes and shapes of the plots of land below. Which plot is the most suitable for your skyscraper? To find out more about each plot, answer these questions.

1) Find the area of each plot in square feet.

2) How much larger is the area of
   a) Plot B than Plot C?
   b) Plot E than Plot B?

3) Only two plots have exactly the same perimeter. Which two plots are they?

PLOT A — 200 feet × 200 feet

PLOT B — 360 feet × 260 feet

PLOT C — 300 feet × 300 feet

PLOT D — 260 feet × 140 feet

PLOT E — 330 feet × 260 feet, 200 feet

10

Big machines called excavators clear the ground at a building site before construction begins.

Math Activities

The construction clipboards have math activities for you to try. Get your pencil, ruler, and notebook (for figuring out problems and listing answers).

NEED HELP?

- **If you are not sure how to do some of the math problems, turn to pages 28 and 29, where you will find lots of tips to help get you started.**

- **Turn to pages 30 and 31 to check your answers. (Try all the activities and challenges before you look at the answers.)**

- **Turn to page 32 for definitions of some words and terms used in this book.**

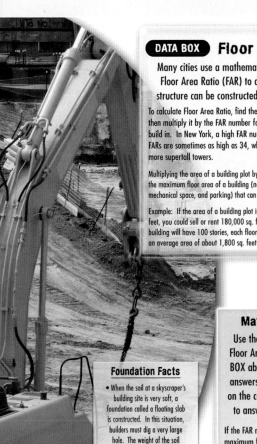

DATA BOX **Floor Area Ratio**

Many cities use a mathematical formula called Floor Area Ratio (FAR) to decide how large a structure can be constructed on a building plot.

To calculate Floor Area Ratio, find the area of the building plot, then multiply it by the FAR number for the city you want to build in. In New York, a high FAR number is 18. In Chicago, FARs are sometimes as high as 34, which is why Chicago has more supertall towers.

Multiplying the area of a building plot by a FAR number determines the maximum floor area of a building (not including basements, roofs, mechanical space, and parking) that can be sold or leased in that city.

Example: If the area of a building plot in New York is 10,000 square feet, you could sell or rent 180,000 sq. feet (10,000 x 18). If the building will have 100 stories, each floor of the building could have an average area of about 1,800 sq. feet (180,000 ÷ 100).

Foundation Facts
- When the soil at a skyscraper's building site is very soft, a foundation called a floating slab is constructed. In this situation, builders must dig a very large hole. The weight of the soil removed must equal the weight of the building so that the final load on the soil does not change and the building stands firm.
- Skyscrapers are very heavy! Taipei 101 is said to weigh about 770,000 tons, which is more than 1.5 billion pounds.

Math Challenge

Use the information about Floor Area Ratio in the DATA BOX above, along with your answers to the first question on the construction clipboard, to answer these questions.

If the FAR number is 15, what is the maximum floor area that could be sold or leased for a building on

1) Plot A?
2) Plot B?
3) Plot C?
4) Plot D?
5) Plot E?

11

Math Facts and Data

To complete some of the math activities, you will need information from a DATA BOX, which looks like this.

Math Challenge

Green boxes, like this one, have extra math questions to challenge you. Give them a try!

You will find lots of amazing details about skyscrapers in FACT boxes that look like this.

GOING UP!

You have been asked to design and build a skyscraper for a multinational bank. The bank wants the structure to be at least 1,300 feet tall, and it must be able to withstand high winds and earthquakes. There will be a lot of interest in your skyscraper, so you have to think about its shape as well as its size. Before you start the project, you will need to know about other skyscrapers in the world — their heights, their shapes, and how long they have been standing. Some amazing structures are being planned and built these days. When it is completed in 2008, Burj Dubai, a 160-story skyscraper in the United Arab Emirates, will be the tallest building in the world — but who knows for how long!

Construction File

In the DATA BOX on page 7, you will see facts about some of the tallest skyscrapers in the world. Use this information to help you answer these questions.

1) Which skyscraper on the list
   a) is the tallest?
   b) is the oldest?
   c) has the most stories?
2) How many more stories has
   a) Taipei 101 than CITIC Plaza?
   b) Sears Tower than either of the Petronas Towers?
   c) the Jin Mao Building than Bank of China Tower?
   d) Sears Tower than Shun Hing Square?
   e) the Empire State Building than Two International Finance Centre?
3) How much taller is
   a) either Petronas Tower than Sears Tower?
   b) CITIC Plaza than Central Plaza?
   c) Taipei 101 than Sears Tower?
   d) Two International Finance Centre than CITIC Plaza ?
   e) Shun Hing Square than Bank of China Tower?

Two International Finance Centre

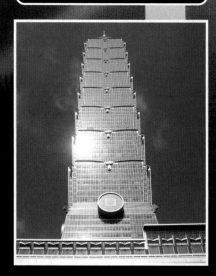

Taipei 101

The World's Tallest Skyscrapers

Name	Location	Year Completed	Number of Stories	Height in Feet
Bank of China Tower	Hong Kong, China	1989	70	1,209
Central Plaza	Hong Kong, China	1992	78	1,227
Empire State Building	New York, New York, USA	1931	102	1,250
Jin Mao Building	Shanghai, China	1999	88	1,380
Petronas Towers 1 and 2	Kuala Lumpur, Malaysia	1998	88	1,483
Sears Tower	Chicago, Illinois, USA	1974	110	1,450
Shun Hing Square	Shenzhen, China	1996	69	1,260
CITIC Plaza	Guangzhou, China	1997	80	1,283
Taipei 101	Taipei, Taiwan	2004	101	1,670
Two International Finance Centre	Hong Kong, China	2003	88	1,362

Many cities in the United States, such as Atlanta, Georgia, can be recognized by the shapes of the skyscrapers on their skylines.

Spire Facts

- Taipei 101 and the Petronas Towers are considered the world's tallest buildings, but they get much of their height from spires. If you do not count spires, the Sears Tower is the world's tallest building.
  - The Empire State Building was constructed with amazing speed. It took only eighteen months to build. Its spire was intended to be a place for tying up airships.
  - The Chrysler Building hid its spire as long as possible because it was competing with another skyscraper for record height. At the last minute, the spire was jacked up from inside the building.

Math Challenge

Use the information in the DATA BOX above to help you answer these questions.

1) How many of the skyscrapers in the DATA BOX are in China?

2) What is the answer to question 1 as a fraction of the total number of skyscrapers in the DATA BOX?

3) What is the fraction in its simplest form?

4) What is the fraction as a percentage?

RECORD-BREAKING SKYSCRAPERS

Before you build your skyscraper, you should know some of the history of skyscrapers. The word "skyscraper" was first used in the eighteenth century to mean a high-flying flag on a ship. The first time the word was used to describe a building was in the 1880s. Although many tall structures were built for special purposes in the past, the Home Insurance Building, built in Chicago in 1885, is usually considered the first skyscraper because it was the first building completely supported by a steel frame, and it had elevators. At 138 feet, it held the record as the world's tallest building until 1890.

Construction File

The DATA BOX on page 9 contains a graph showing the heights of the tallest buildings in the world during the last approximately seventy-five years. Use the graph to help you answer these questions.

1) About how tall is
   a) the Chrysler Building?
   b) Sears Tower?

2) Which building is about
   a) 1,480 feet tall?
   b) 1,250 feet tall?
   c) 1,670 feet tall?

3) Which building was the tallest in the world in
   a) 1980?   d) 1970?
   b) 2000?   e) 2005?
   c) 1960?

Tall Buildings in History

- People began constructing tall buildings thousands of years ago. Between 2600 BC and 2570 BC, the Red Pyramid in Egypt, at 344 feet tall, held the record as the tallest building in the world.
- The Great Pyramid, built in Giza, Egypt, was 135 feet taller than the Red Pyramid and held the record as the world's tallest building for nearly 4,000 years, from 2570 BC until AD 1300! Between 2570 BC and AD 1439, the Great Pyramid eroded and lost approximately 23 feet of its height.
- Lincoln Cathedral, in England, was completed in 1300. With its wooden spire, it raised the record height for the tallest building to 525 feet. Lincoln Cathedral held the record for more than two centuries, until its spire collapsed in 1549.

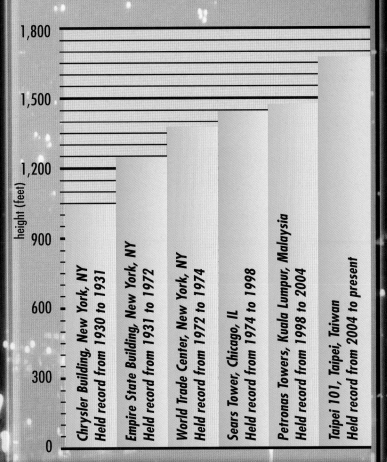

The World's Tallest Building

The graph below shows the approximate heights of the record-holders between 1930 and 2006.

height (feet)

1,800

1,500

1,200

900

600

300

0

Chrysler Building, New York, NY
Held record from 1930 to 1931

Empire State Building, New York, NY
Held record from 1931 to 1972

World Trade Center, New York, NY
Held record from 1972 to 1974

Sears Tower, Chicago, IL
Held record from 1974 to 1998

Petronas Towers, Kuala Lumpur, Malaysia
Held record from 1998 to 2004

Taipei 101, Taipei, Taiwan
Held record from 2004 to present

Math Challenge

In 1930, the Chrysler Building was the tallest building in the world. How many years after 1930 did the following buildings become the world's tallest?

1) the Petronas Towers   4) World Trade Center
2) Empire State Building   5) Sears Tower
3) Taipei 101

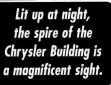

Lit up at night, the spire of the Chrysler Building is a magnificent sight.

THE BUILDING SITE

To design a skyscraper, you must first know the size and shape of the building site, which is the plot of land you will build on. The dimensions of the plot will help you decide the size and shape of your building. Many cities have rules about how large a building can be constructed on a certain size building plot. In some cities, the size of a building has to be based on a formula known as Floor Area Ratio (FAR). In cities that do not use FAR, the size of a building usually must be approved by city planners before construction can begin. After choosing a suitable building site, you are ready to begin your design.

Construction File

Look at the sizes and shapes of the plots of land below.
Which plot is the most suitable for your skyscraper?
To find out more about each plot, answer these questions.

1) Find the area of each plot in square feet.

2) How much larger is the area of
   a) Plot B than Plot C?
   b) Plot E than Plot B?

3) Only two plots have exactly the same perimeter.
   Which two plots are they?

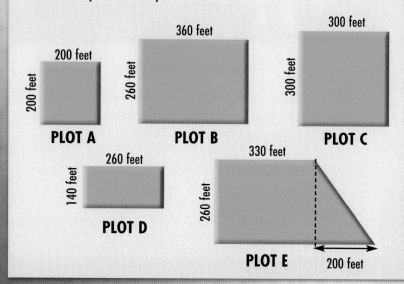

PLOT A — 200 feet × 200 feet

PLOT B — 360 feet × 260 feet

PLOT C — 300 feet × 300 feet

PLOT D — 260 feet × 140 feet

PLOT E — 330 feet, 260 feet, 200 feet

Big machines called excavators clear the ground at a building site before construction begins.

10

11

A SKYSCRAPER'S STRENGTH

When you have selected a plot of land and know the size of the floor area for your skyscraper, you must start thinking about the building materials you will use. For a structure this size, you need very strong materials. You should use reinforced concrete, which is concrete with steel bars inside. It is very strong, and the concrete helps keep the steel bars from weakening in a fire. The whole structure should have a steel skeleton, built much like putting together a giant construction set. Because wood and masonry have limited strength, they can be used only for buildings of about fifteen stories or less.

Construction File

The shapes below are the basic dimensions, or "footprints," of some buildings.

Find the exact perimeters of these buildings, then round your answers to the nearest 10 feet.

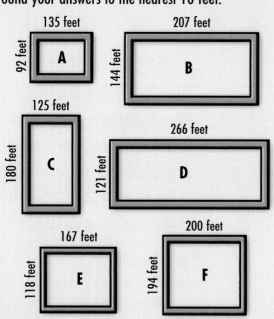

A — 135 feet × 92 feet
B — 207 feet × 144 feet
C — 125 feet × 180 feet
D — 266 feet × 121 feet
E — 167 feet × 118 feet
F — 200 feet × 194 feet

Math Challenge

A building has a perimeter of 600 feet.

What are the length and width of the building if its area is

1) 22,500 sq. feet?   4) 12,500 sq. feet?
2) 20,000 sq. feet?   5) 16,875 sq. feet?
3) 17,600 sq. feet?

Concrete Facts

• New concrete usually takes about one month to gain its full strength.
• Concrete is very strong in resisting compression, or pushing, loads, but it is weak in resisting tensile, or pulling, loads.

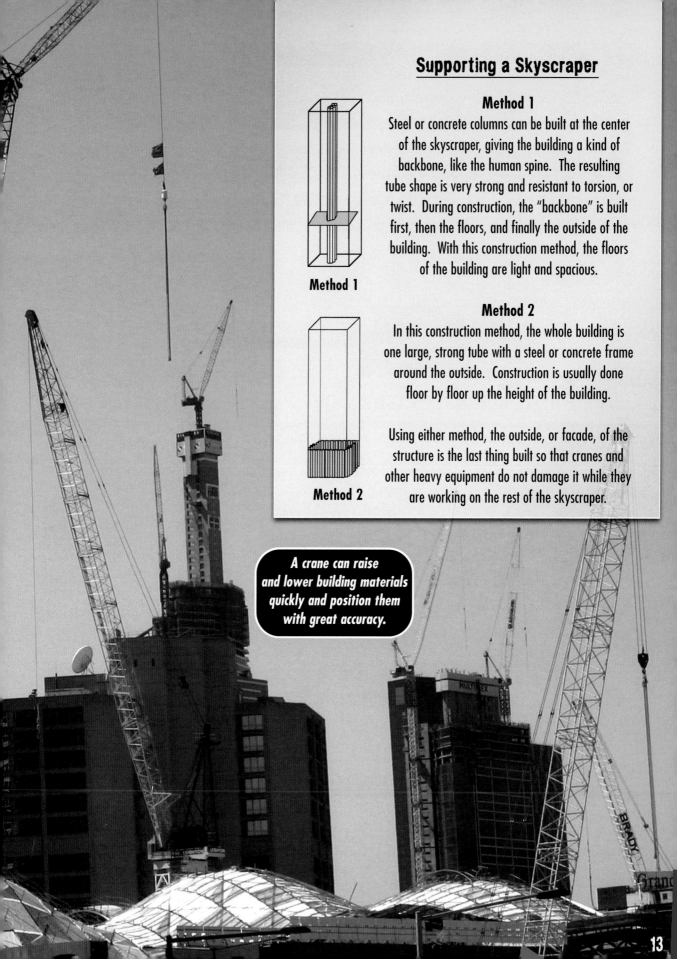

Supporting a Skyscraper

Method 1

Steel or concrete columns can be built at the center of the skyscraper, giving the building a kind of backbone, like the human spine. The resulting tube shape is very strong and resistant to torsion, or twist. During construction, the "backbone" is built first, then the floors, and finally the outside of the building. With this construction method, the floors of the building are light and spacious.

Method 1

Method 2

In this construction method, the whole building is one large, strong tube with a steel or concrete frame around the outside. Construction is usually done floor by floor up the height of the building.

Method 2

Using either method, the outside, or facade, of the structure is the last thing built so that cranes and other heavy equipment do not damage it while they are working on the rest of the skyscraper.

A crane can raise and lower building materials quickly and position them with great accuracy.

THE DESIGN

Skyscrapers come in all shapes and sizes. Since a structure is usually bolted together like a huge construction set, the only real limits on shape and size are the imaginations of the architects and engineers who design the building and put the pieces together. What shape would you like your skyscraper to be? What do you need to think about to decide? The amount of money you have is one factor. You also need to know the purpose of the skyscraper. Will it be used for offices or homes, or a mixture of both? Will any part of it be used for shops or as a hotel? How will the use of the building affect its design?

Construction File

Here are some building designs of different shapes. Name all of the shapes that make up each building.

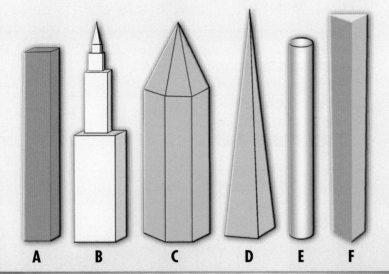

A B C D E F

Design Fact

Skyscrapers can be built for offices or for residences, which are places where people live. If the building is for residences, designers must make sure that all living areas have or are near windows. People want natural light coming into their homes and enjoy looking out at a view. If the skyscraper is for offices, having every area close to a window is less important. The use of artificial light is common in all areas of office buildings.

Design Fact

It takes a team of many architects and engineers to design a skyscraper. The team for the Burj Dubai grew to seventy people during the design phase. A team may spend one to two years designing a skyscraper.

Math Challenge
Which of the following cuboid buildings has the greatest volume?

Building A: 165 feet long, 100 feet wide, 1,300 feet tall

Building B: 130 feet long, 65 feet wide, 1,475 tall

Building C: 200 feet long, 130 feet wide, 1,150 feet tall

As this view of the Chicago skyline shows, the shape of most skyscrapers is a rectangular prism, or cuboid.

A GOOD FOUNDATION

When constructing a tall building, you must make sure it is on solid ground. If you build on soil that is too soft, the structure will lean or sink, and it might even fall down. It is important to dig down under any soft soil to the hard ground underneath. Even after reaching solid ground, a builder will pour in lots of reinforced concrete to make a foundation even firmer and to spread the weight of the building evenly. In soft soil, deep foundations called piles or caissons (pronounced KAY-sahns) are used to support skyscrapers. Piles are like stilts or pillars that are planted in the strong, solid rock found deep underground.

To make a foundation that is strong enough for a skyscraper, workers build a network of steel bars and pour concrete over them.

Construction File

The diagram in the DATA BOX on page 17 represents the lower floors of a tall building. Use the information it provides to answer these questions.

Which floor is
1) one level below ground floor 0?
2) four levels above basement −2?
3) five levels below floor 3?
4) six levels above basement −1?
5) three levels below basement −1?
6) nine levels above floor 5?
7) fifteen levels above basement −3?
8) nine levels below floor 15?

Floor Fact

Not all skyscrapers have a thirteenth floor. Because the number 13 is often considered unlucky, the floors in some buildings skip from 12 to 14. In China, the number 4 is unlucky, so buildings there often skip from floor 3 to floor 5.

Foundation Facts

- Pile foundations are a lot like thick stilts that reach down to firm ground to support the weight of a tall building.
- Where firm ground does not exist, or when it is too deep to reach with piles, a slab foundation might be used to spread the weight, or load, of the building over the surrounding soil. If the load is spread evenly, the soil under the slab will support the weight of the building.
- When a building is very heavy, a combination of pile and slab foundations can be used to share the load between them.

building with pile foundation   **building with slab foundation**   **building with pile and slab foundation**

Math Challenge

An elevator is traveling up and down the building shown in the DATA BOX to the right.

Where will the elevator end up after this series of movements?
- starts on ground floor 0
- goes up three floors
- goes down six floors
- goes up nine floors
- goes up two floors
- goes down four floors
- goes down five floors

DATA BOX

Floor-to-Floor

The floor diagram below represents only the lower part of a skyscraper. The building actually has eighty-eight floors above ground level and four basement levels below ground.

floor 16
floor 15
floor 14
floor 13
floor 12
floor 11
floor 10
floor 9
floor 8
floor 7
floor 6
floor 5
floor 4
floor 3
floor 2
floor 1
ground floor 0
basement −1
basement −2
basement −3
basement −4

ground level

SUPPORTING THE STRUCTURE

The overall weight of a skyscraper is supported by its "skeleton," or frame, but each floor also needs to be supported. This support comes from concrete and steel beams and girders. Concrete columns or walls can also provide support. Architects and building engineers think carefully about the shapes that will be formed when girders, columns, and concrete walls are joined together. Some shapes, especially triangles, are much stronger than others. Architects can be as imaginative and decorative as they like when designing the outside of a building, but engineers must get the structure right so the building will be safe.

Construction File

Triangles are used to design strong structures. The triangles below are three different types.

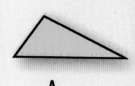
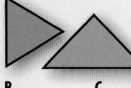

A   B   C

D   E   F

1) Which of these triangles is
   a) scalene?   b) isosceles?   c) equilateral?
2) Which of these triangles have a right angle?

Math Challenge

The DATA BOX on page 19 contains information about the weight, per foot, of a girder.

How heavy is a girder that is
1) 13 feet long?   4) 46 feet long?
2) 23 feet long?   5) 82 feet long?
3) 36 feet long?

Column Facts

• Steel support columns come in many different shapes. They can be I-shaped, like girders, or they can be box-shaped or circular. Concrete columns can also be many different shapes.
• The support columns for many tall buildings are made with both steel and concrete. A steel I-shaped column might be encased within a larger concrete column, or a steel box-shaped column might be filled with concrete. Combining these two materials increases the strength and stiffness of a column.

Girders

Girders are commonly shaped like the letter I because that shape does not bend easily. Girders come in different lengths and weigh about 70 pounds per foot of length.

The Strength of Shapes

Unlike many other shapes, a triangle is very strong. The strength of a triangle comes directly from its shape. In fact, a triangle cannot be changed unless one of its joints or sides breaks.

Other shapes can be changed with downward pressure.

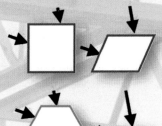

To make a rectangle or a square stronger, a diagonal strut is used to turn the shape into two triangles. The strut keeps the rectangle or square from being pushed flat or sideways.

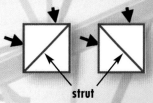

strut

The frame of a skyscraper is a series of triangles. A triangle is the strongest shape.

Construction Fact

A large team of construction workers is responsible for building a skyscraper. Some construction workers operate the cranes that move and place materials. Others mix and pour concrete. Workers also connect the steel beams, girders, and columns together. The construction of a skyscraper can easily take two or three years.

RESISTING EARTHQUAKES

Making sure that a skyscraper can withstand earthquakes is very important, particularly in certain parts of the world. The building will need a stiff frame made of beams, columns, and concrete walls, with triangular steel braces for extra strength. A strong frame will keep a skyscraper safe in both high winds and earthquakes. Wind pushes on a building from above the ground, while earthquakes push on the building from below the ground. During an earthquake, the ground moves. If the building does not move with it, the building will fall down. Modern skyscrapers are strong and flexible. They are built to survive earthquakes that will damage many smaller buildings.

Construction File

The DATA BOX on page 21 describes how the Richter scale is used to measure earthquakes.

The following numbers stand for different sizes of earthquakes. Put them in order of size from the smallest tremor to the most serious quake.

2.4   6.3   3.0   1.7   5.4   7.2   5.0   8.3

The shape of this San Francisco skyscraper helps it withstand earthquakes.

Math Challenge

1) Look at the decimals below. Where would you place each decimal on the line?

2.8   3.5   3.2   2.4   2.5   3.0   2.2   3.9

```
├──┼──┼──┼──┼──┼──┼──┼──┼──┼──┼──┼──┼──┼──┼──┤
2                     3                      4
```

2) Now round each of the decimals to the nearest whole number.

Swaying Facts

• Early skyscrapers, such as the Empire State Building, were built to be very solid. Even in extreme winds, these buildings sway only about 20 inches at the top.

• Newer skyscrapers, which are much taller than earlier skyscrapers, are built with more flexible materials and are designed to sway up to 5 feet at the top in extreme winds.

When earthquakes occur in cities, older buildings are often damaged, while most skyscrapers remain safe.

Quake Fact

Large earthquakes can last about 10 to 45 seconds. They are often followed by one or more smaller quakes, called aftershocks. The aftershocks can cause some damaged buildings to collapse completely.

The Richter Scale

12 — The ground shakes heavily and becomes distorted, causing major damage or complete destruction, with objects thrown into the air.

11 — Few buildings remain standing; bridges and railroads are destroyed; and water, gas, electricity, and telephones are out of action.

10 — The ground cracks open, many buildings are destroyed, and some landslides occur.

9 — A huge earthquake, causing major damage and loss of life over an area of more than 600 miles.

8 — A great earthquake, causing great destruction and loss of life over an area of about 200 miles.

7 — A major earthquake, causing serious damage over an area of about 60 miles.

6 — Trees sway and buildings are damaged within an area of about 30 miles. Some damage is caused by collapsing and falling objects.

5 — Movement feels like a heavy truck striking a building. People sleeping are awakened.

4 — People can feel movement, and objects shake.

3 — Some people feel vibrations, like those caused by heavy traffic.

2 — Only very sensitive people notice movement.

1 — Microquakes usually not felt by people are recorded on local seismographs, which are machines that measure earthquakes.

ELEVATORS AND STAIRWAYS

When building a skyscraper of one hundred or more floors, it is vital to think about how people will get up and down inside the building. Not many people are willing to climb one hundred flights of stairs, except perhaps in an emergency. You need to put in elevators, but you must also think about safety. If there is a fire or a power failure, using elevators will be unsafe. The building will need stairs as well. Where will you put emergency stairways? How many stairways will you put in? Also useful in emergencies are helicopter landing pads, which many tall buildings now have on their roofs.

Construction File

A skyscraper has 100 stories above ground level. Each flight of stairs between the stories has 22 steps.

1) How many steps would a person have to climb from the ground floor to reach floor
   - a) 3?
   - b) 4?
   - c) 10?
   - d) 16?
   - e) 45?
   - f) 50?
   - g) 70?
   - h) 100?

2) Each step is about 6 inches high. About how high would a 22-step flight of stairs be?

3) Use the answer to question 2 to help you estimate the height, in feet, of 100 flights of stairs.

Math Challenge

Use the information in the DATA BOX on page 23 to help you answer these questions.

1) How long will it take a standard elevator to travel, without stopping, from ground level to
   - a) floor 10?
   - b) floor 25?
   - c) floor 60?
   - d) floor 100?

2) About how long will it take an express elevator to travel, without stopping, from ground level to
   - a) floor 10?
   - b) floor 25?
   - c) floor 60?
   - d) floor 100?

3) How much faster is the express elevator than the standard elevator for each of the trips above?

Emergency Facts

People are always trying to design better systems for evacuating people from high-rise buildings quickly and safely in emergencies. Means of evacuation can include external telescopic ladders and stairs, rope devices for people to tie around their waists and lower themselves down the outside of a building, and even emergency escape tubes. Many of these ideas, however, are not suitable for elderly people or children — or for anyone in a very tall building! Every city has regulations that specify the emergency safety equipment that must be included in a new building.

DATA BOX ## Elevator Speeds

The distance between the floors, or floor-to-floor height, of the skyscraper you are designing is 10 feet.

Your standard elevators travel at about 20 feet per second.
Your express elevators travel at about 33 feet per second.

The designs of emergency stairways are usually very plain. Most people, after all, will never use them.

Elisha Otis

Commercial skyscrapers were not possible until 1853, when American inventor Elisha Otis developed a device to make elevators in tall buildings safe. The device was an emergency brake, which prevented elevators from free-falling if their mechanisms failed. Otis's braking system gave people confidence that elevators were practical and safe to use. The first "modern" elevators were steam-powered. Today's elevators are hydraulic or electric.

THE BUILDING PROCESS

Your designs have been approved by architects and engineers, and you must now start building the skyscraper. Construction might take several years. You will need lots of skilled people to help you. First, you need people who specialize in digging foundations. Later, you will need construction workers. When the building is finished, you will need electricians, plumbers, and plasterers. Because building a skyscraper can be dangerous, you also need to think carefully about the safety of the workers. Be sure that everyone on the building site wears a hard hat and steel-toed boots for protection.

Construction File

It generally takes one week to build each floor of a skyscraper.

Use the calendar on the right to help you answer these questions.

1) If the building work for the first floor starts on Feb 12, when will the
   a) second floor be completed?
   b) fourth floor be completed?
   c) seventh floor be completed?
   d) ninth floor be completed?
   e) sixteenth floor be completed?
   f) twentieth floor be completed?

2) If the building work starts on Feb 12, estimate how long it will take to complete the
   a) fifty-second floor.
   b) one-hundred-fourth floor.

3) Which floor is being built on
   a) February 18?
   b) February 27?
   c) March 20?
   d) May 1?

FEBRUARY

	1	2	3	4	5	6
7	8	9	10	11	12	13
14	15	16	17	18	19	20
21	22	23	24	25	26	27
28	29					

MARCH

		1	2	3	4	5
6	7	8	9	10	11	12
13	14	15	16	17	18	19
20	21	22	23	24	25	26
27	28	29	30	31		

APRIL

					1	2
3	4	5	6	7	8	9
10	11	12	13	14	15	16
17	18	19	20	21	22	23
24	25	26	27	28	29	30

MAY

1	2	3	4	5	6	7
8	9	10	11	12	13	14
15	16	17	18	19	20	21
22	23	24	25	26	27	28
29	30	31				

JUNE

			1	2	3	4
5	6	7	8	9	10	11
12	13	14	15	16	17	18
19	20	21	22	23	24	25
26	27	28	29	30		

The site manager is in charge of a large team of workers, a lot of machinery, and all of the building materials.

Safety Facts

Building a skyscraper is dangerous, but, today, there are many rules to protect workers.
- Before the outside walls are constructed, the floors of the building must be fenced to prevent workers from falling.
- Hard hats and steel-toed boots must always be worn on the construction site.
- When there are high winds, construction usually must stop because working conditions are too dangerous.
- When climbing on beams and girders, workers must use safety ropes that tie them to the building.

Skyscraper Who's Who

Architects and engineers work together to design a skyscraper.
- Architects decide what the outside of the building will look like and how it will fit into the surrounding landscape. They also design the layout of the inside space, such as how the rooms are divided and arranged and where the elevators, hallways, and restrooms will be.
- Engineers make sure the building is strong enough to support its own weight as well as the weight of all the people, furniture, and other things that will be inside it. They also ensure that the building is properly heated and cooled and that water and electricity are available where needed. Above all, engineers must be sure the building will be safe in strong winds and earthquakes.

Math Challenge

A skyscraper took 3 ½ years to design and build.

How long is this time period in
1) months?
2) weeks?
3) days?

YOUR FINISHED SKYSCRAPER

Congratulations! You're finished! Your skyscraper is built. As you stand on the ground and look up, you think it is the most impressive building in the world. Other people seem to agree with you, and the offices and apartments in the skyscraper have all been leased. Within the building, the finishing touches have been made to the rooms, and most of the furnishings are in place. Before the building can be used, however, you have to get approval from the safety inspectors. After the building is approved, people can start working and living in it. You hope they will enjoy your skyscraper.

Construction File

In the DATA BOX on page 27, you will see a table of data for the Sears Tower. Use the data to help you answer these questions.

1) How many windows are there in the whole building?

2) The antennae rise above the normal height of the building. How tall are the antennae?

3) One ton is 2,000 pounds. How many pounds of steel were used to build the Sears Tower?

4) What is the area of the base of the Sears Tower?

Sears Tower

Coin Fact

Have you ever heard that dropping a coin from the top of a tall building could kill a person on the ground below? The statement is not true. The coin could hurt someone, but the injury is very unlikely to be fatal. A coin dropped from the tallest skyscraper is traveling at the same speed when it hits the ground as a coin dropped from a four-story building because, when an object falls a long distance, it reaches a point where air resistance, which slows down a falling object, is equal to the pull of gravity. At that point, the object stops gaining speed and will not fall any faster.

Sway Fact

On the upper floors of skyscrapers, chandeliers and water in tubs or glasses are sometimes seen swaying along with the building in high winds. Although the motion of the building causing this phenomenon is very safe, people who notice the movement often feel uncomfortable or unsettled.

The Sears Tower

The table below contains information about
the Sears Tower in Chicago, Illinois.

height	1,450 feet
height at the top of the antennae	1,730 feet
weight of steel used	83,750 tons
other materials used	concrete, aluminum, glass
foundation type	piles
year completed	1974
number of floors	110
number of windows on each floor	146
number of elevators	106
base measurements	230 feet x 230 feet

Your skyscraper is attracting a lot of attention. It will soon become a recognized landmark in the city, just like these impressive skyscrapers in Denver, Colorado.

Math Challenge

The tops of tall, modern buildings sway from side to side in strong winds. The heights of the skyscrapers below are measured in inches. To estimate the maximum amount that each building might sway in a very strong wind, divide the height of the building by 500.

How many inches might the top of each of these skyscrapers sway?

1) 20,000 inches   3) 14,800 inches   5) 15,400 inches   7) 16,800 inches
2) 17,600 inches   4) 16,600 inches   6) 15,600 inches   8) 15,000 inches

MATH TIPS

PAGES 6–7

Construction File

TOP TIP: When subtracting numbers with lots of digits, make sure that you line up all the digits so the units (ones) line up with the units, the tens with the tens, the hundreds with the hundreds, and so on.

PAGES 8–9

Math Challenge

TOP TIP: To find the number of years between two dates, subtract the older date from the newer date.

Example:
```
   1974
 − 1931
     43
```
so 1974 is 43 years after 1931

PAGES 10–11

Construction File

The area of a shape is the amount of space inside the shape. Area is measured in square units, such as square inches or square feet.

To find the area of a rectangle, multiply its length by its width.

8 x 4 = 32 sq. inches

To find the area of a triangle, multiply its base by its height and divide by 2.

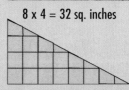

8 x 4 = 32 ÷ 2 = 16 sq. inches

The perimeter is the distance all the way around the edge of a shape. To find the perimeter of a rectangle, add its length and its width, then double the answer.

8 + 4 = 12 x 2 = 24 inches

Math Challenge

A quick way to multiply a number by 15 is to, first, multiply the number by 10, then divide that answer by 2 and add the two answers together.

PAGES 12–13

TOP TIP: When rounding numbers to the nearest 10, remember that numbers ending in 5, 6, 7, 8, or 9 round up to the next multiple of 10, and numbers ending in 1, 2, 3, or 4 round down to the previous multiple of 10.

Math Challenge

You know that the length plus the width will be half the perimeter (300 feet), so choose pairs of numbers that add up to 300, such as 200 and 100, 80 and 220, and so on. Multiply the two numbers in each pair to find the area.

PAGES 14–15

Math Challenge

The volume of a three-dimensional shape, such as a rectangular prism, or cuboid, is the measure of the space inside the shape. The volume is measured in cube units, such as cubic inches or cubic feet. To find the volume of a cuboid, multiply its length by its width by its height.

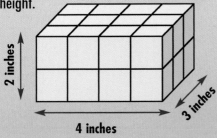

2 inches

3 inches

4 inches

Volume = 4 x 3 x 2 = 24 cubic inches

PAGES 18–19

Math Challenge

When multiplying a number by 70, first multiply the number by 7, then multiply that answer by 10.

Example: 76 x 70 = 76 x 7 = 532 x 10 = 5,320

Remember that to multiply a number by 10, move each digit of the number one place to the left and use a zero to fill in the empty column.

Example:
532 x 10 = 5,320

Tth	Th	H	T	U
		5	3	2
	5	3	2	0

PAGES 20–21

Math Challenge

TOP TIP: When rounding decimals to the nearest whole number, remember that a decimal of 5, 6, 7, 8, or 9 will round up to the next whole number, and a decimal of 1, 2, 3, or 4 will round down to the previous whole number.

PAGES 22–23

Math Challenge

To figure out how long each elevator will take to travel between the ground floor and a higher floor, first find the distance the elevator has to travel for each trip by multiplying the number of floors by the floor-to-floor height.

After you find the distance, divide by 20 for the standard elevator and divide by 33 for the express elevator.

PAGES 24–25

Math Challenge

Remember that each year has 12 months, 52 weeks, and 365 days (366 days in a leap year). You can estimate the number of days in $3\frac{1}{2}$ years by using the number of weeks. There are 7 days in each week, so multiply the number of days in a week by the number of weeks in $3\frac{1}{2}$ years.

PAGES 26–27

Math Challenge

To divide a number by 500, divide first by 1,000, then double your answer.

To divide a number by 1,000, move each digit of the number three places to the right.

Example: 16,800 ÷ 1,000 = 16.8

Tth	Th	H	T	U	.	t
1	6	8	0	0		
	1	6	8	0	.	0
		1	6	8	.	0
			1	6	.	8

ANSWERS

PAGES 6–7

Construction File

1) a) Taipei 101
   b) Empire State Building
   c) Sears Tower

2) a) 21
   b) 22
   c) 18
   d) 41
   e) 14

3) a) 33 feet
   b) 56 feet
   c) 220 feet
   d) 79 feet
   e) 51 feet

Math Challenge

1) 6   2) $\frac{6}{10}$ 3) $\frac{3}{5}$ 4) 60 percent (60%)

PAGES 8–9

Construction File

1) a) 1,050 feet
   b) 1,450 feet

2) a) Petronas Towers
   b) Empire State Building
   c) Taipei 101

3) a) Sears Tower
   b) Petronas Towers
   c) Empire State Building
   d) Empire State Building
   e) Taipei 101

Math Challenge

1) 68 years
2) 1 year
3) 74 years
4) 42 years
5) 44 years

PAGES 10–11

Construction File

1) A = 40,000 sq. feet   D = 36,400 sq. feet
   B = 93,600 sq. feet   E = 111,800 sq. feet
   C = 90,000 sq. feet

2) a) 3,600 sq. feet   3) A and D
   b) 18,200 sq. feet

Math Challenge

1) 600,000 sq. feet   4) 546,000 sq. feet
2) 1,404,000 sq. feet   5) 1,677,000 sq. feet
3) 1,350,000 sq. feet

PAGES 12–13

Construction File

A = 454 feet rounds to 450 feet
B = 702 feet rounds to 700 feet
C = 610 feet (no rounding needed)
D = 774 feet rounds to 770 feet
E = 570 feet (no rounding needed)
F = 788 feet rounds to 790 feet

Math Challenge

1) 150 x 150   3) 220 x 80   5) 225 x 75
2) 200 x 100   4) 250 x 50

PAGES 14–15

Construction File

A) a rectangular prism, or cuboid
B) three rectangular prisms, a cube, and
   a square-based pyramid
C) a hexagonal prism and a hexagonal-based pyramid
D) a square-based pyramid
E) a cylinder
F) a triangular prism

Math Challenge

Building C (29,900,000 cubic feet) has a greater volume than Building A (21,450,000 cubic feet) or Building B (12,463,750 cubic feet).

PAGES 16–17

Construction File

1) basement −1
2) floor 2
3) basement −2
4) floor 5
5) basement −4
6) floor 14
7) floor 12
8) floor 6

Math Challenge

basement −1

PAGES 18–19

Construction File

1) a) A and D
   b) C and E
   c) B and F

2) C and D each have one right angle

Math Challenge

1) 910 pounds
2) 1,610 pounds
3) 2,520 pounds
4) 3,220 pounds
5) 5,740 pounds

PAGES 20–21

Construction File

1.7   2.4   3.0   5.0   5.4   6.3   7.2   8.3

Math Challenge

1)

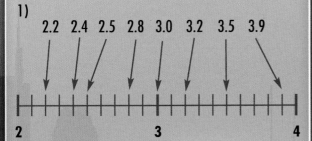

2.2   2.4   2.5   2.8   3.0   3.2   3.5   3.9

2   3   4

2) 2.8 rounds to 3
   3.5 rounds to 4
   3.2 rounds to 3
   2.4 rounds to 2
   2.5 rounds to 3
   3.0 rounds to 3
   2.2 rounds to 2
   3.9 rounds to 4

PAGES 22–23

Construction File

1) a) 66   e) 990   2) 132 inches (11 feet)
   b) 88   f) 1,100   3) 1,100 feet
   c) 220   g) 1,540
   d) 352   h) 2,200

PAGES 22–23 (continued)

Math Challenge

1) a) 5 seconds   2) a) 3 seconds   3) a) 2 seconds
   b) 12½ seconds   b) 7½ seconds   b) 5 seconds
   c) 30 seconds   c) 18 seconds   c) 12 seconds
   d) 50 seconds   d) 30 seconds   d) 20 seconds

PAGES 24–25

Construction File

1) a) February 25   d) April 14
   b) March 10   e) June 2
   c) March 31   f) June 30

2) a) 1 year
   b) 2 years

3) a) first floor   c) sixth floor
   b) third floor   d) twelfth floor

Math Challenge

1) 42 months
2) about 182 weeks
3) 1,274 to 1,278 days

PAGES 26–27

Construction File

1) 16,060 windows
2) 280 feet
3) 167,500,000 pounds
4) 52,900 sq. feet

Math Challenge

1) 40 inches   5) 30.8 inches
2) 35.2 inches   6) 31.2 inches
3) 29.6 inches   7) 33.6 inches
4) 33.2 inches   8) 30 inches

GLOSSARY

BRACES steel girders placed on an incline against walls or floors to strengthen a structure

CAISSONS structural supports for the foundation of a building, drilled deep into earth and often embedded in bedrock

COLUMNS vertical pillars that support the weight of a building

COMPRESSION the process of pressing or pushing down on material, usually reducing its size

CUBOID a rectangular prism with the approximate shape of a cube

DIMENSIONS measures of size, such as length, width, and height

EQUILATERAL TRIANGLE a triangle with all three sides of equal length

EVACUATING emptying or moving out of a place in an organized way, especially to escape harm or some kind of danger

FACADE the outer front or face of a building

FLOOR AREA RATIO (FAR) a formula that some cities use to control the sizes of tall buildings

FOOTPRINTS outlines of the areas of ground or surface space covered by buildings or other structures

FRAME the skeletonlike structure that provides the main support for a building

GIRDERS large, main beams made of wood or steel that provide structural support, especially in ceilings and floors

HYDRAULIC moved or operated by means of water or some other liquid

ISOSCELES TRIANGLE a triangle that has two sides of equal length

LEASED rented, instead of owned

LOAD weight, especially the total weight of a building

PILE FOUNDATIONS stiltlike structures made of concrete that support a building below the ground

PLOT an area of land available for building some kind of structure on

RATIO a comparison of two quantities, one divided by the other

RICHTER SCALE a numerical scale that measures the magnitude of earthquakes

SCALENE TRIANGLE a triangle with all three sides of unequal length

SEISMOGRAPHS devices that measure and record earthquakes and other vibrations in the earth

SKYLINES the outlines of buildings against the sky or the horizon

SLAB FOUNDATION a flat concrete structure that supports a building at or below ground level

SPIRE the tapered roof or spike at the top of a skyscraper

STORY one level, or floor, of a building

STRUT a brace that adds strength to a frame by resisting compression along its length

TENSILE able to withstand tension, or pulling

TORSION a twisting force

Measurement Conversions

1 inch = 2.54 centimeters (cm)

1 square inch = 6.4516 square centimeters (cm^2)

1 cubic inch = 16.39 cubic centimeters (cm^3)

1 foot = 0.3048 meter (m)

1 square foot = 0.0929 square meter (m^2)

1 cubic foot = 0.0283 cubic meter (m^3)

1 mile = 1.609 kilometers (km)

1 pound = 0.4536 kilogram (kg)

1 ton = 0.9074 tonne